DIG!

To Harlow Shapley

DIG!

A Journey under the Earth's Crust

Written and illustrated by
John and Faith Hubley

Harcourt Brace Jovanovich, Inc.
New York

The authors wish to acknowledge the aid and assistance of
Dr. Bruce C. Heezen, Associate Professor of Geology at Columbia University,
who provided much useful information and who read the work in manuscript.

Adapted from the film Dig! by John and Faith Hubley
Copyright © 1973 by John and Faith Hubley
Design Supervision: Kay Susmann
All rights reserved. No part of this publication may be reproduced
or transmitted in any form or by any means, electronic or mechanical, including
photocopy, recording, or any information storage and retrieval system,
without permission in writing from the publisher.
Lyrics to "Rocco's Song," "Aunt Sadie's Song," "Uncle Iggy's Song,"
and "Cousin Meta's Song" by John and Faith Hubley and Quincy Jones,
Copyright © 1972 by Ulla Music.
Printed in the United States of America
First edition

BCDEFGHIJK

Library of Congress Cataloging in Publication Data: Hubley,
John. Dig! A journey under the earth's crust. "Adapted from the
film Dig!" SUMMARY: A boy and his dog travel in geological
time guided by a talking rock. [1. Geology—Fiction] I. Hubley,
Faith, joint author. II. Dig! A journey into the earth. [Motion picture] PZ7.H8634Di3 [Fic] 72-88169 ISBN 0-15-223490-X

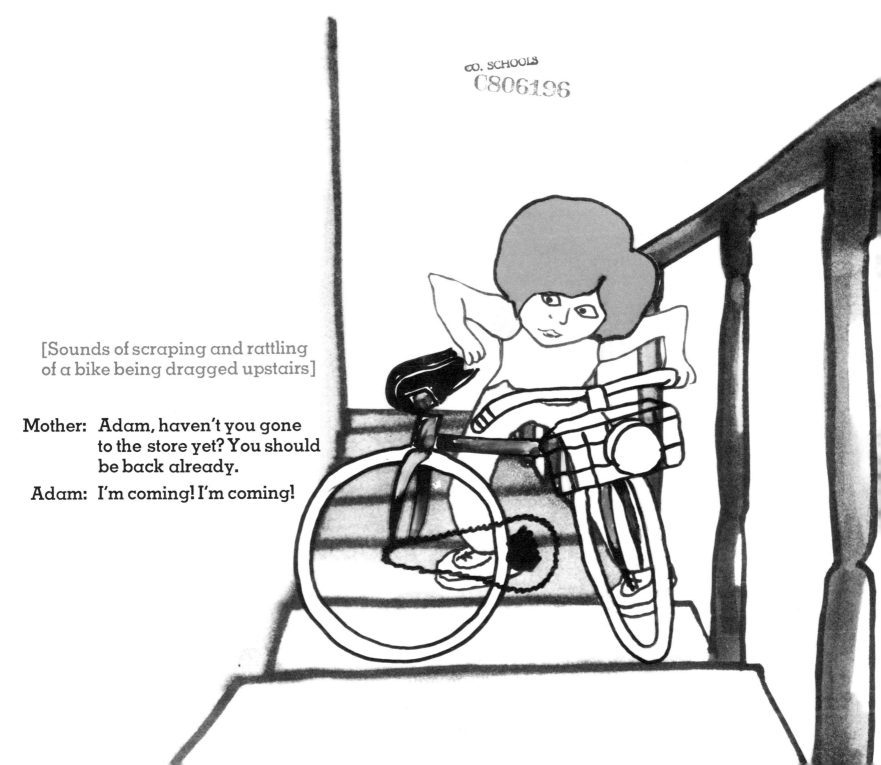

Adam: What am I supposed to get again?

Mother: One quart of milk!

Mother: Do you have to take your dog?

Adam: I promised him. Come on, Bones!

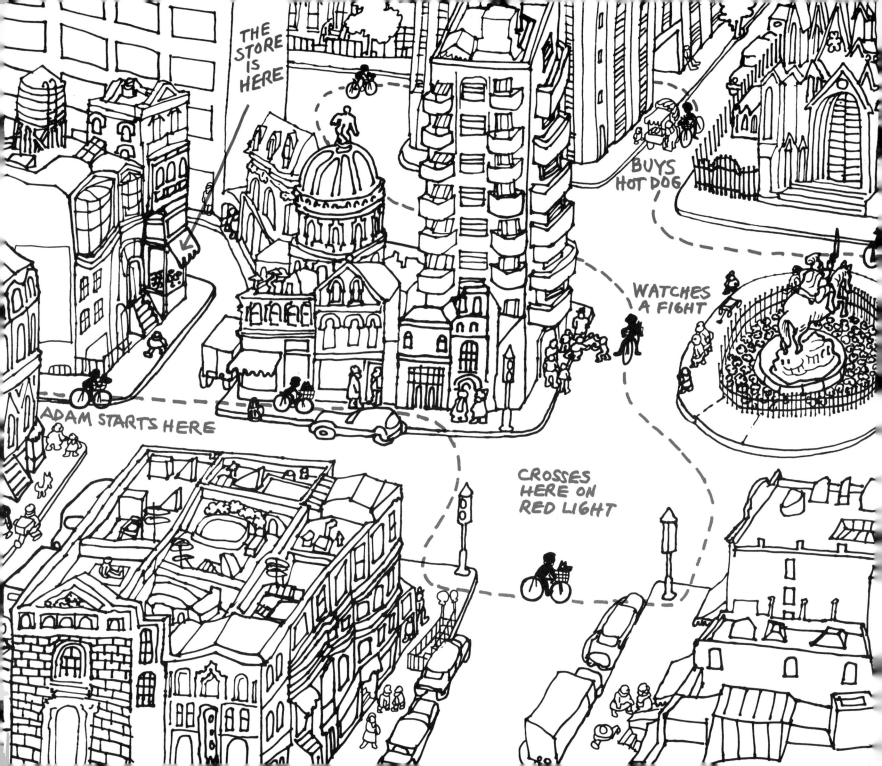

Adam: Hey, Bones! That crane has hooked my bike!

Adam: It's falling on those rocks!

Rocco: Psst. Hey, kid!
Adam: What?

Rocco: Do me a favor, kid. Lend me your bike. I've got to get going.

Adam: Going where?

Rocco: Home—down that hole. Take a look!

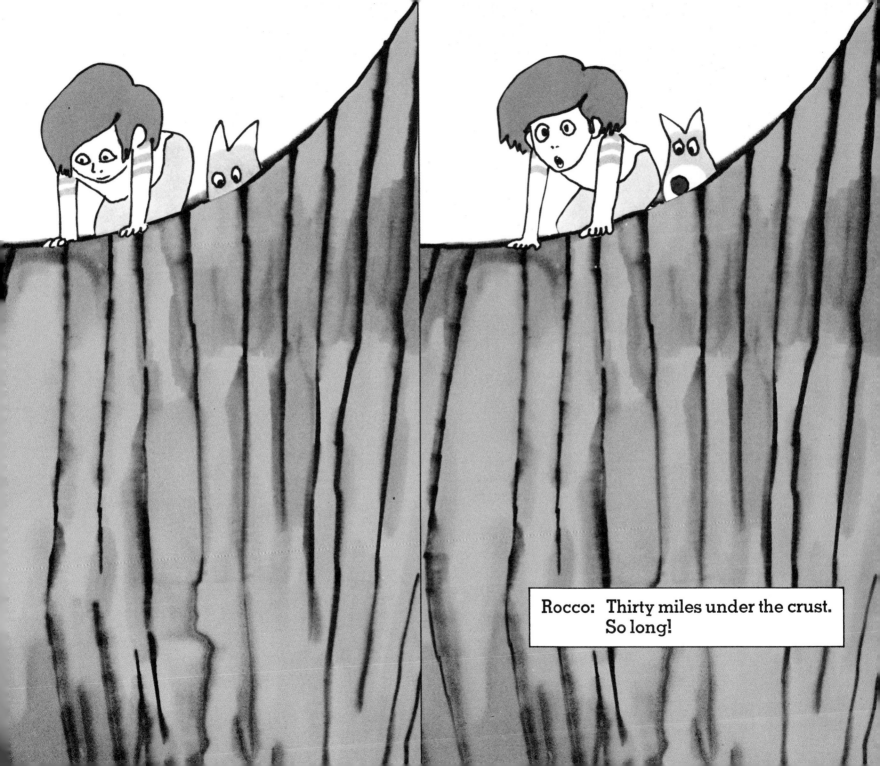

Adam: He's going down that hole on my bike!

[Sound of Adam falling deep into the hole landing with a splash]

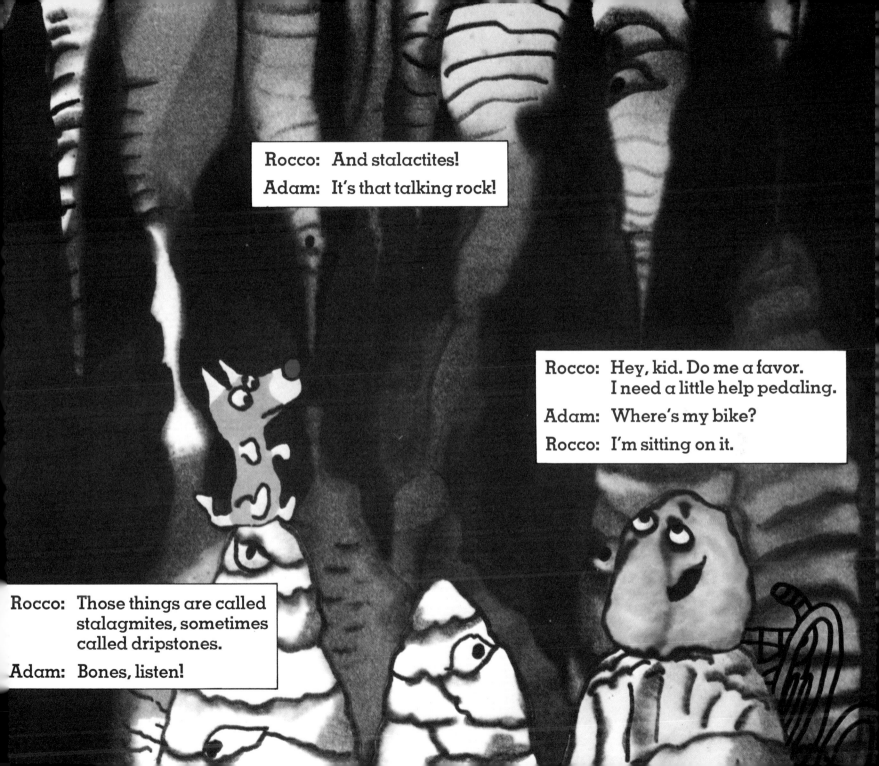

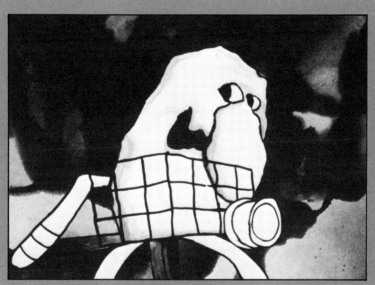

ROCCO'S SONG
Just call me Rocco the rover
with a heart made of gold,
and I ramble like a rolling stone.

I know a shortcut out,
a little rounabout.
It's a secret hidden trail of my own. Yeah!

Sh, I think I hear Mother Nature making stew.
She's brewin' gobs and bogs of hot gassy goo.

She's gonna flip her lid.
You better watch it, kid.
There she blows!

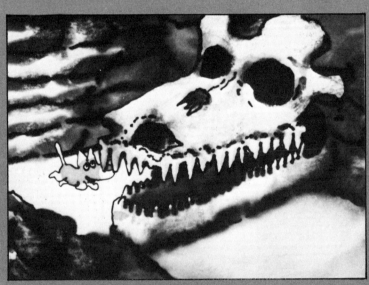 

Betcha nickel to a dime
I can guess where Father Time
stashed the bones of a dusty dinosaur. Yeah!

We take a midnight ride down the rock-bottom road,
bumpty bumpty bump bump bump.
You got nothing to fear, kid,
as long as I'm near.
Just keep moving those pedals
and I'll steer you clear.
Low bridge! Down the rock-bottom road!
I can tell by your breath
that you're scared half to death,
but I bet you'd never guess where we are.
Now if the earth is like an apple,
we're riding in the skin,
and we still haven't traveled very far.
Oh, the midnight ride down the rock-bottom road,
bumpty bumpty bump bump bump.

[Pressure sounds]

Rocco: Do you feel the earth tremble? Do you feel a tingle of the ground waves?

Adam: We're going sideways!

Rocco: Hang on, they'll be coming faster! I love it!

Adam: You love it, but I can't steer!

Rocco: It's not my fault. It's the earth's fault. Fault of the earth! Heh, heh! A fault is a crack in the crystal rocks!

Adam: What's that rumble?

Rocco: That's the shakers, kid, shoving and pushing, pushing and shoving. The big rock layers are slipping—look up there! We're in an earthquake!

Adam: How often do these earthquakes happen?

Rocco: About a million times a year.

Adam: What are those shiny lights?

Rocco: Salt crystals are forming.

Adam: Regular salt?

Rocco: Sure! Salt is the most plentiful mineral on earth.

Rocco: There goes your dog, kid. You'd better get him. Once the crystals start to grow, you can't stop them.

Adam: Stop barking, Bones. Bones! It's only you!

Adam: Are we almost home?
Rocco: We're getting there, kid.

Rocco: This canyon looks familiar. I used to live on this strata!

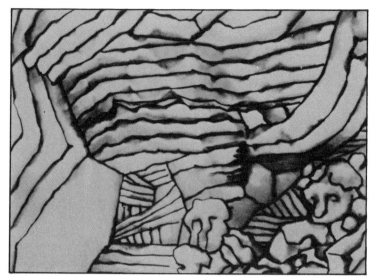

Rocco: My folks! My old rock tribe! I haven't seen them for ten million years!

Rock Tribe: Rocco! What a surprise! You haven't changed a bit!

Rocco: Adam, say hello to my folks—Uncle Iggy, Aunt Sadie, Cousin Meta.

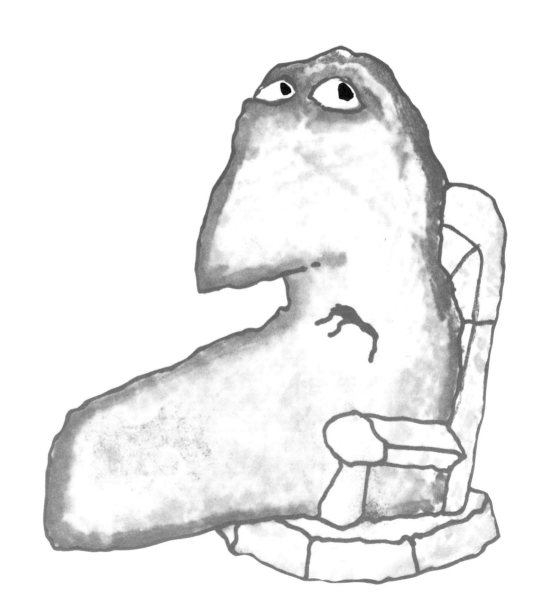

Adam: Hello. This is my dog, Bones.

Rocco: How's it been going, Uncle Igneous?

Iggy: Hard. It's hard to be a rock!

IGGY'S SONG

I was born in a river of fire,
I was weaned on steam and heat,
I burned my way through solid rock
To cheat the crowded street.
I pushed and shoved to bust the crust,
to meet the mountaintop.
The blast was heard for miles around—
zoom plop!

Now I was stone cold up on that mountain,
wind and rain beating holes in my head.
Old Man River came and carried me home.
Now I'm back in my underground bed.
Uncle Iggy Iggy Iggy Igneous is my name.
Taught Rocco all he knows and that's the truth.
I'm pushy as a groundhog
and hard as a possum's tooth.
Iggy Iggy Iggy Iggy Iggy Uncle Igneous,
Uncle Igneous is my name.

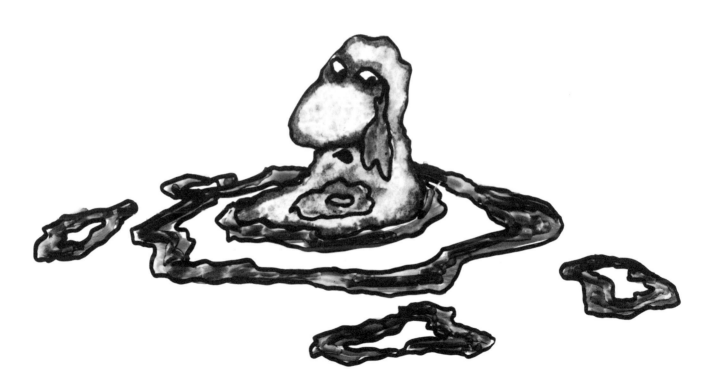

Rocco: Aunt Sadie's got a song about being sedimentary!

SADIE'S SONG
Troubles of a sedimentary rock

Rocks in my head,
mud in my eyes,
bones in my bed,
rain makes me cry.
Don't talk about rivers—
they give me the shivers.
Seldom have I seen a sadder Sadie
so sedimentary.

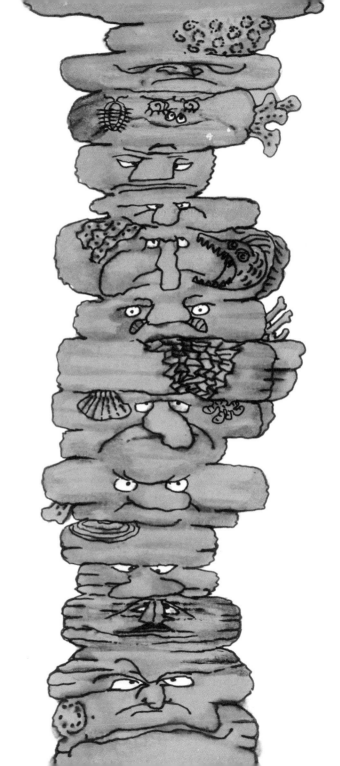

Dirt in my ear,
sand on my back,
year after year,
oh, another crack!
Don't talk to me about rivers—
they give me the shivers.
Seldom have I seen a sadder Sadie.
Oh, ho ho ho!

Rocco: Cousin Meta's got a song too!

META'S SONG

I'm your Cousin Meta-morphic.
How'm I doing?
Almost porfic.

If you'd rather, call me Meta.
I can change from what I am to
something betta.

First I'm a limestone.
I keep changing into marble.

I keep changing
when I'm sandstone,
I keep changing into quartz.

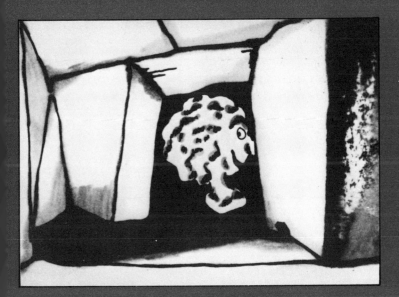

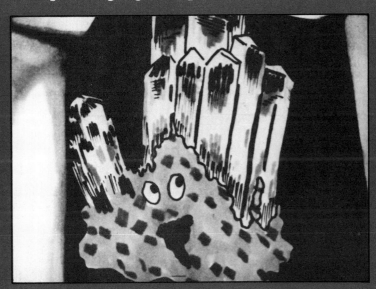

I do all sorts of rearranging.
Many times I've changed and made
me into jet or jade.

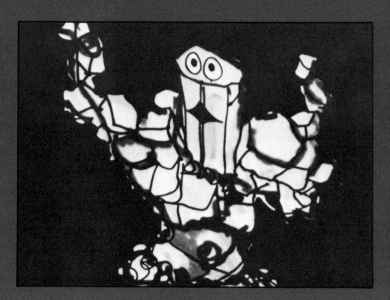

I'm switching constantly of late
from certain kinds of shale to slate.
Uncle Iggy and Aunt Sadie

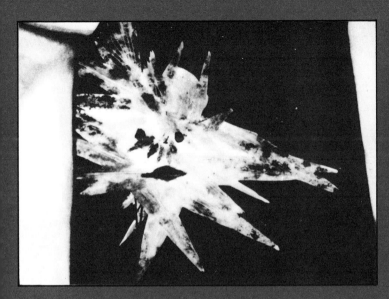

sometimes wonder if I'll ever be a lady,
but I tell them Cousin Meta
always changes into something that is betta.

Sniff! Sniff!

Adam: Rocco, Rocco! I can't find Bones.

Rocco: So long, folks! I gotta look for a dog.

Adam: He must have sniffed a bone or something!

Rocco: Quick! That's it! To the fossil boneyard!

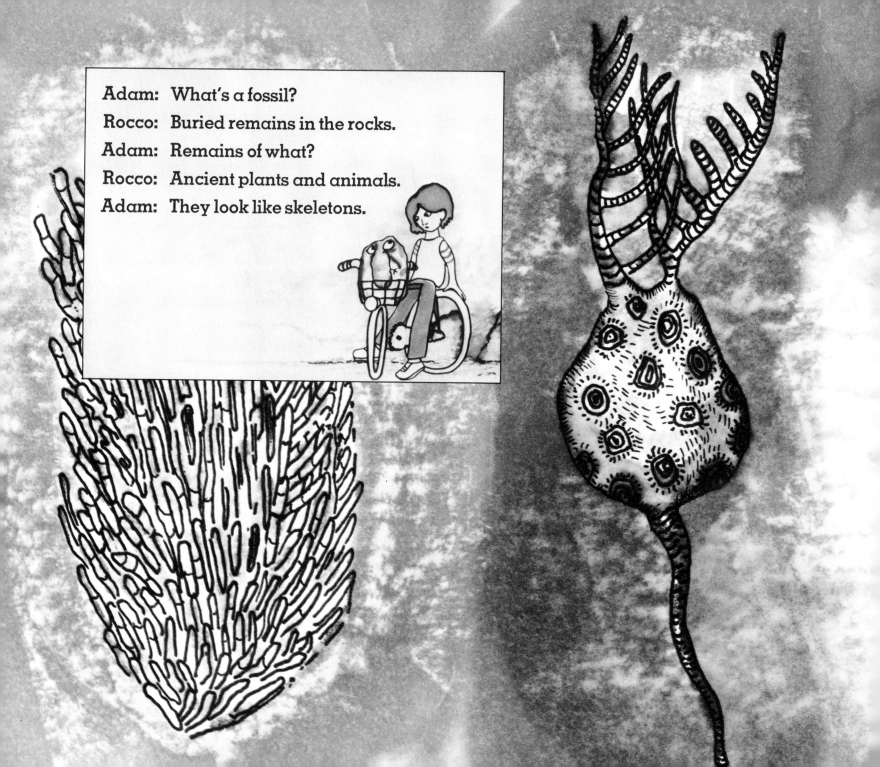

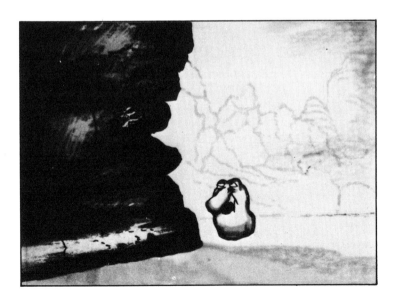

Fossil: Speak of endless time! Speak of fossiled canyon walls! Of bones of ancient ages! The mysterious birth of planet earth!

Rocco: Have you seen a little gray dog?

Fossil: He speaks to me of dogs. Very well. Speaking of dogs—the gray one slid back into the whirling sands of time. He's lost in search of bones!

Adam: Lost! I've got to find him! This way?

Fossil: Not so. Turn around. You too must go backward, backward to the very beginnings, back to the early hours of earth time.

Rocco: I'm not going back to any beginning.

Adam: Yes you are! Be strong—you're a rock!

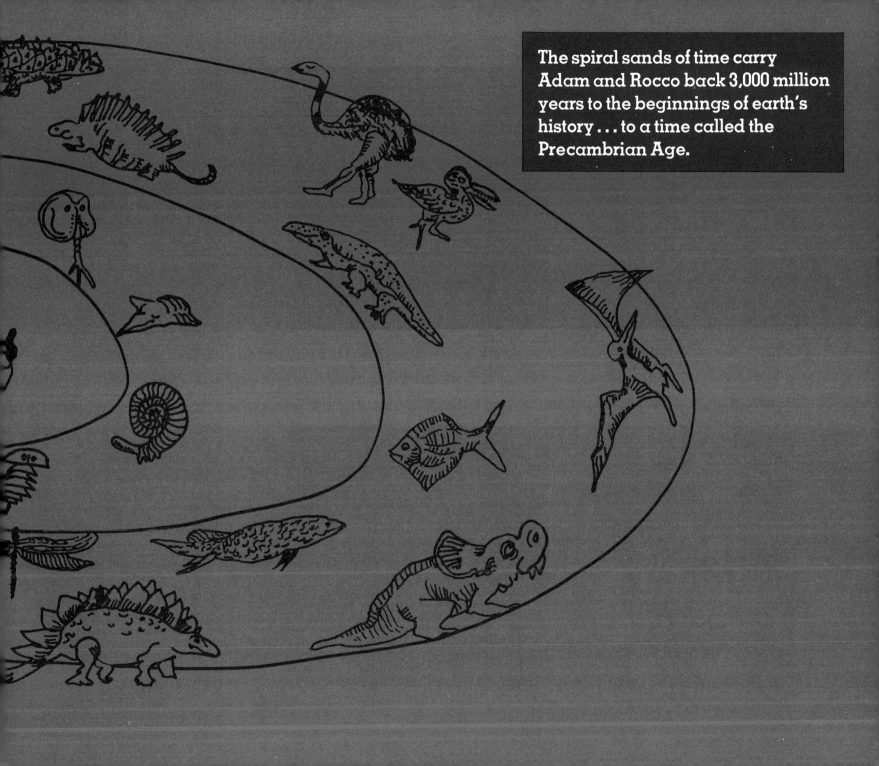

The spiral sands of time carry Adam and Rocco back 3,000 million years to the beginnings of earth's history... to a time called the Precambrian Age.

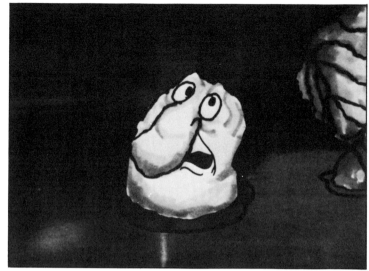

Rocco: I give up! This is too far back.

Rocco: This is probably before there were any bones.

Adam: There's something prehistoric-looking.

Rocco: They look like Eozoic rocks. They're among the oldest. I once knew one 3 billion years old.

Rocco: Can you tell me where we are, sir?

First Rock: Bev domig zeb!

Rocco: I can't understand! Maybe Bones is in the next age. Let's go!

Adam: Look! Those rocks are pushing up!

Rocco: Rising mountains. That's called uplift. Could be the Adirondacks. Come on. If we're lucky, we'll get a ride to where life began.

Rocco: Ah ha! This is the Paleozoic Age!

Adam: Paleo-zo what?

Rocco: Paleo-zo-ick. Lots of water— and lots of fish. Life began in the water.

Adam: I think I hear Bones!

Rocco: What's the matter with that animal?

Adam: He must have sniffed a bone or something.

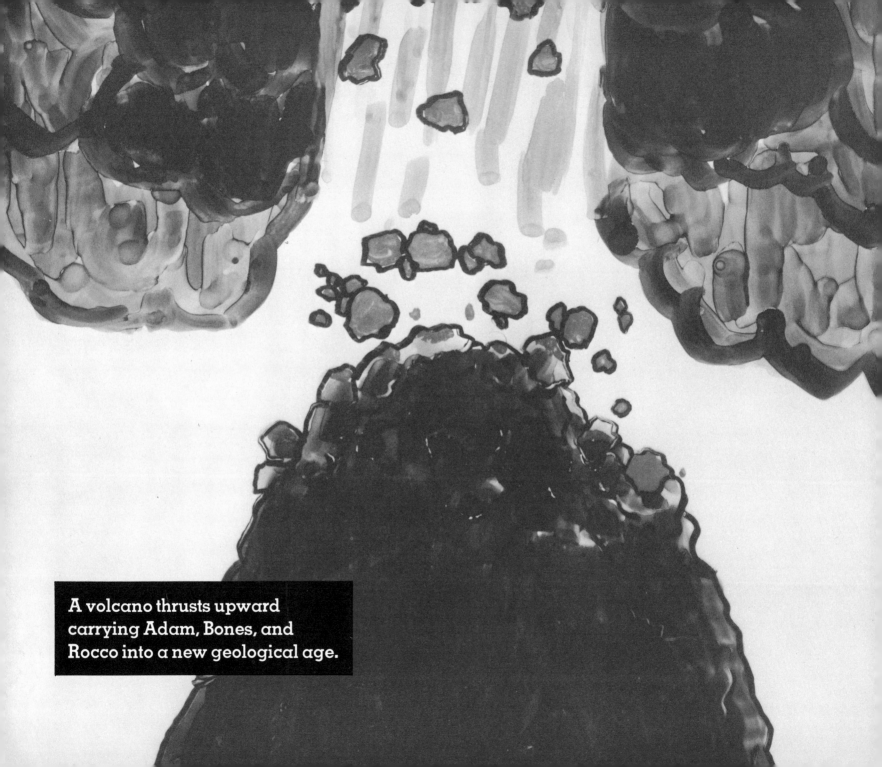

A volcano thrusts upward carrying Adam, Bones, and Rocco into a new geological age.

Adam: Hear that! He's over there!

Rocco: Now he's chewing on fossil rocks! Don't you ever feed that dog?

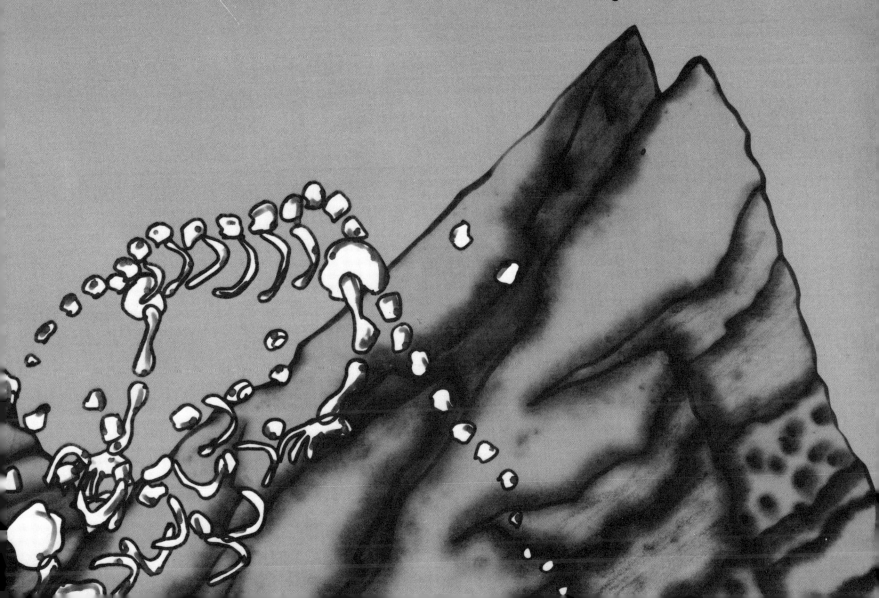

Adam: What's that?

Rocco: Get on your bike! Grab Bones! The Ice Age is coming! You gotta migrate!

Adam: I'm freezing! The rivers are chasing us! The ocean is sinking!

Rocco: Faster, Adam! The forests are going next. You've got to catch up with the present! You can't survive the cold!

The huge glacier tumbles with a booming crash.

Adam: Bones, we're back in the present. Hear them blasting rock?

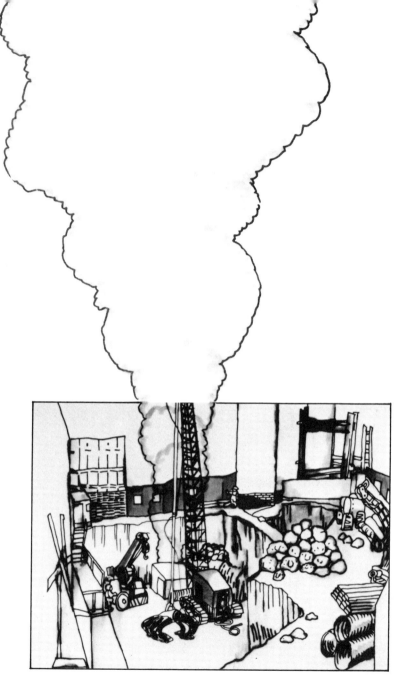

Adam: It's late... we're goin' home!

Kid: Your father's lookin' for you.

Mother: Adam, where have you been? Dinner's ready. Did you bring the milk?

Adam: Sorry, Mom. I forgot.

Mother: Can't you remember anything?

Adam: I remember stalagmite—stalactite—igneous—

Adam: Sedimentary—Metamorphic—Paleozoic—Mesozoic—Cenozoic!!

And Adam really begins to think about our good planet earth.